Lights, Camera, Action!

Secrets to Master Your On-Camera Presence Like a PRO

WORKBOOK

By Judy Go Wong

Special Thanks

I want to thank my wonderful parents for capturing countless cherished family and professional moments in photographs throughout my life. Without their influence, I would not be where I am today.

Thank you to the talented photographers who have taken my headshots over the past decade. Not only did they provide me with exceptional photos, but they also gave me valuable insights into the art of posing, sharing tips to avoid double chins and neck wrinkles. Working with these skilled professionals was a wonderful experience, and I appreciate the education I gained from them.

I am grateful for the CEOs, entrepreneurs, and individuals I have connected with on social media and through webinars who face the challenge of presenting themselves on camera. Their experiences have inspired me to help them enhance their appearance, confidence, and skills. It's a privilege to help them shine on camera!

Thanks to my KBA coach, Dr. Chonta Haynes, who has been with me since this workbook journey began. I am grateful for her wisdom, guidance, and unwavering patience throughout the process. Her support means the world to me.

My heartfelt thanks go to my incredible friends who proof-read this manuscript and supported me throughout this process. A special shout-out to David Moser, Rick Quan, Travis Reaves, Evelyn Moy (Family Association and Community Leader), Linda Nelson-Garrett, and Jackie Lee for catching typos and offering brilliant suggestions that made this workbook even better for readers.

The Journey of Judy Go Wong

Judy Go Wong is a SAG-AFTRA actor, award-winning film-maker, and seasoned certified coach passionate about helping others shine on camera and in person. Over the past decade, Judy has appeared in notable TV series such as *House of Cards* and *VEEP*, demonstrating her expertise in on-camera presence.

As a multifaceted creative, Judy has worked in every aspect of production, including acting, producing, directing, screenwriting, set design, MUA (makeup artist) and wardrobe styling. She is the author of "*How to Start a Film Acting Career Right Where You Live.*" She has coached many actors, filmmakers, and digital creators to refine their on-camera skills.

Judy has mentored young filmmakers at the Smithsonian's Freer Sackler Teen Council, teaching life skills and building self-esteem. She also served as the Director of Operations for Asian Pacific American Film, Inc. Today, Judy coaches CEOs, entrepreneurs, actors, filmmakers, and digital creators, helping them master their on-camera presence to boost confidence and professionalism.

With years of experience in creative and coaching, Judy dedicates herself to helping others shine on camera and in their professional lives.

Turning Up the Lights!

You are beautifully and wonderfully made. Embrace looking fabulous and bring joy to the world. It takes effort to present your best self, and you should appreciate your efforts.

For the past decade, I've been coaching actors. Helping them get great headshots and frame themselves for self-tape (self-tape is an industry phrase for videoing yourself) auditions. They've learned how to pose and feel more confident. I now share my skills with CEOs, entrepreneurs, filmmakers and digital creators.

As social media continues to flourish, there's a growing need for captivating headshots and engaging self-taped videos.

No matter your job, taking care of your appearance is essential. The right hairstyle, makeup, smile, posture, and outfit can help you stand out and feel more confident. Mastering the basics of creating great self-taped videos is essential because you only get one chance to make a lasting first impression.

I'm so excited for you to embark on this journey. Are you ready? Let's go!

Table of Contents

Chapter 1

Posing Like a Pro

Whether striking a confident pose or capturing a candid moment, there are endless possibilities to explore when presenting yourself in front of the camera. It involves experimenting and finding your unique style and personality, and if you're an actor, the roles you want to play will all influence this fun and valuable skill.

I find posing entertaining — perhaps I've enjoyed it from the beginning. My mom would deck me out like Shirley Temple when I was a preschooler. One day, my family and I were walking in a department store, and our destination was its massive basement cafeteria, a big thing back in the 1960s. With so many of us, slipping away unnoticed was a cinch. I'd zip ahead, hiding amidst the clothing racks, then pose next to the mannequins on their lofty platform on the corner. But here's the kicker: my family strolled right by me without a clue! I couldn't help but giggle when I overheard my brother asking where I was. Mom's bewildered response, a classic, was: "Judy, nei hia bin doh ah?" ("Where are you?"), she called out in Cantonese.

Leave it to my sharp-eyed older brother to spot me up there with the mannequins. Mom's laughter echoed through the store as she marveled at my perfect stillness. "Judy, you make one fantastic mannequin!" she chuckled. With a grin, I hopped down and joined them, relishing the joy of entertaining my family. And that's how my posing adventures began — I just loved that feeling of blending in like a real-life mannequin.

Here are some posing actions you might find helpful.

Keep your body relaxed and avoid stiffness to look your best in photos. This will give you a more natural and attractive pose that you will love. So be yourself and let your inner beauty shine through. Be sure to maintain good posture. It can have a positive impact on your well-being and enhance your self-assurance.

It is more beneficial to angle your body (to the side of your choice) instead of facing the camera straight on. This technique can improve your overall appearance and produce more flattering results. Experiment with different angles before a full-length mirror to find what works best.

Are you aware that some individuals have a preferred side of their face when being photographed? Experiment with various positions and angles to find your most flattering side, then take some photos and review them to determine which is most suitable for you. Discovering your best angle can enhance your appearance and boost your confidence while smiling and posing in front of the camera.

Here's a pro tip to help you rock your next photoshoot: Don't hang your arms straight at your sides. Instead, use them to add personality and interest to your pose. This will create a dynamic presence that radiates confidence and competence. Directions are below.

FOR THE GALS: If you want a more dynamic and engaging pose, distribute your weight on one foot by adopting a "hip stance." To achieve this pose, imagine a clock lying flat on the ground, and you're standing right in the center of the clock.

If your left side is more flattering:

1. Stand with both feet together.

2. Bring your left leg and foot about six inches forward.

3. Have your left foot point it at 12:00.

4. Adjust your left knee forward, then slightly tilt it to the right.

5. Your right leg should hold most of the weight.

6. Right foot, point it at 2:00.

7. Place your left hand atop your left hip.

8. Bring your left arm back, just a little.

9. Keep your four fingers together visible in front.

10. Your thumb is not visible in the back of your waist.

11. Keep your right arm and hand on the side. (Turn your hand so your thumb is visible and facing forward.)

If the right side is more flattering:

1. Stand with both feet together.

2. Bring your right foot and leg about six inches forward.

3. Right foot, point it at 12:00.

4. Adjust your right knee forward, then tilt the knee to the left a little.

5. Your left leg should hold most of the weight.

6. Left foot, point it at 10:00.

7. Place your right hand atop your right hip.

8. Bring your right arm back, just a little.

9. Keep your four fingers together visible in front.

10. Your thumb is not visible at the back of your waist.

11. Keep your left arm and hand on the side. (Turn your hand so that your thumb is visible and facing forward.)

This stance exudes confidence, poise, and elegance in formal settings, such as business or academic environments and in all your photos.

FOR THE GUYS: If you want a more dynamic and engaging pose, distribute your weight on your feet. To achieve this pose, imagine a clock lying flat on the ground, and you're standing right in the center of the clock.

If the left side is more flattering:

1. Stand straight.

2. Have your feet aligned, three inches apart.

3. Bring your left foot three inches forward, pointing at 11:00.

4. The right foot stays in the same spot. Point your right foot at 1:00.

5. Adjust your left knee forward, this causes you to pivot to the right.

6. Most of your weight should be on the right leg.

7. If you are wearing a sports coat or suit jacket, position your left hand in your left pocket.

8. Placing four fingers in the pocket of your sports coat or suit jacket, leaving your thumb out. You can also try it with your entire hand in your "pants pocket". Moving your jacket behind your wrist.

9. If you are not wearing a jacket, put four fingers in your "pants pocket" while leaving your thumb out.

10. Then tilt your head a little to the right and smile, then you'll be in an exceptional pose.

If the right side is more flattering:

1. Stand straight.

2. Have your feet aligned, three inches apart.

3. Bring your right foot three inches forward, pointing at 1:00.

4. The left foot stays in the same spot. Point your right foot at 11:00.

5. Bend your right knee forward, this causes you to pivot to the left.

6. Most of your weight should be on the left leg.

7. If you are wearing a sports coat or suit jacket, position your right hand in your right pocket.

8. Placing four fingers in the pocket of your sports coat or suit jacket, leaving your thumb out. You can also try it with your entire hand in your pants pocket, moving your jacket behind your wrist.

9. If you are not wearing a jacket, put four fingers in your right pants pocket while leaving your thumb out.

10. Tilt your head to the left a little and smile, then you'll be in an exceptional pose.

Follow these steps, and practice them, and you'll look dapper when posing for your photos and/or headshots.

This posture results in a more natural, relaxed, and pleasing appearance and can be effective when presenting. Remember to tilt your head for a more comfortable and friendlier look.

Proper execution of this posture is essential, as it can enhance one's presence and command attention. So, practice, practice, and practice it in front of your full-length mirror until your pose becomes natural and you love the way you look.

It's beneficial to experiment with a range of facial expressions in front of your mirror to communicate and convey emotions. Depending on the context and desired effect, this can include a genuine smile or a more severe, contemplative expression. It is essential to relax the muscles in your face and jaw, avoiding unnecessary tension or strain, to achieve a natural appearance. By taking the time to practice and master different facial expressions, you can enhance your ability to communicate and connect with others in both personal and professional settings, as well as on camera.

It is essential to remain mindful of any tension in your body and release it while you pose. Tension can cause a pose to appear unnatural and forced, detracting from its intended purpose. I recommend practicing awareness and relaxation before a full-length mirror to mitigate stress and keep a natural appearance.

If you want to create depth and avoid a flat appearance in your photographs, position yourself away from the background. In doing so can achieve greater depth and three-dimensionality in your images, making them visually charming.

When engaging in any photographic or video-based activity, it is essential to be mindful of the lighting conditions as well (another chapter later). The quality of illumination can influence the outcome and overall appearance of your pose. It is crucial to experiment with lighting to discover the most favorable conditions that enhance your physical attributes. You should balance adequate lighting with the proper contrast, color temperature, and intensity.

As I have mentioned, practicing in front of a mirror is a great way to improve your posture and facial expressions. This low-stress, fun approach allows you to assess your preferred pose and expressions and find areas that require improvement. Regular practice can enhance your professional demeanor and convey confidence. Remember that "perfect practice makes perfect"!

Your confidence level can affect the success of a pose. Trusting your abilities and believing in yourself to look presentable and professional in photos, on camera, on the red carpet, and in person is imperative. Projecting confidence is key to the desired image.

Take some notes as you practice posing. What's working and what's not?

Let's proceed to the next chapter, "Ready, Set, Go!". We're focusing on the essentials and figuring out what we can leave behind.

Notes

Chapter 2

Ready, Set, Go!

Are you prepared for the camera? Do you often feel uncomfortable or self-conscious when someone wants to take your picture? How about recording yourself on video? Would you rather not step in front of a camera at all? If any of this sounds familiar, you're not alone. Learn simple tips to feel confident and shine in front of a camera. With "Lights, Camera, Action!" you can master on-camera presence like a pro.

It's taken me decades to figure out how to look and feel good on camera, so relax. You can take this journey at your own speed. I'm here to guide you, give precise directions, and lead you through your transformation. Together, we'll ensure you feel great and love how you look.

Let's start with a self-assessment to identify specific areas for improvement that you may notice. Some areas might interest you, while others may not. Before you begin, take out your latest headshots or past photos, like selfies, for your assessment. Skip the sections that aren't relevant to you. If any areas require attention, mark the box with a checkmark. But if everything is going well, leave it blank, and we'll keep cruising!

Let's begin by discussing your hairstyle. Are you satisfied with it? Does it complement the shape of your face? Do you like the color of your hair? Do highlights appeal to you, or would you prefer to dye your hair another color? Please make your notes below under Hairstyle. Chapter 2 includes a discussion on "Hairstyles."

☐

Hairstyle

_____

_____

_____

Look at your chosen makeup colors (if you wear makeup). Do they harmonize with your skin tone? Men, this is for you too. It helps to have concealer if you have dark circles underneath your eyes, foundation to balance your skin tone, and press powder to take the shine away. Also, if you don't have full eyebrows, use eyeshadow or eyebrow mascara (that matches your hair color) to make your eyebrows look fuller. Don't forget about brushes that are appropriate for each item. Please note the items you may need below. In Chapter 3, we will cover "Makeup Magic for Everyone."

☐

Makeup

_____

_____

_____

Review your headshots and past photos, paying attention to your smile. Do you like your smile? If not, please explain how you would like your smile to appear (with a huge grin, with a less pronounced grin, showing your teeth, not showing your teeth).

☐

Smile

_____

_____

_____

Examine the same pictures for indicators of a double chin, a wrinkled neck, poor posture, poses, and wardrobe choices. If any of these concerns you, mark them next to the appropriate items below and check the box. In Chapters 5 through 9, we will cover all the details of these topics.

☐

Double Chin

_____

_____

_____

☐

Wrinkled Face and Neck

_____

_____

_____

☐

Posture

_____

_____

_____

☐

Posing

_____

_____

_____

☐

Wardrobe

_____

_____

_____

To maintain a youthful appearance and address concerns about aging, identify the areas on your face, neck, and upper chest that could benefit from anti-aging serums and creams with the most recent photos of yourself. With your findings, please take notes under "Youthful Look" below.

☐

Youthful Look

_____

_____

_____

Consider your confidence level and whether it reflects in your photos and videos. Do you look professional? Do you look like an expert? Would you feel comfortable using your services based on your appearance? Are you reflecting on your brand? In the "Confidence" section below, please write down your findings. We'll address confidence in Chapter 11.

☐

Confidence

_____

_____

_____

Do you know how to frame yourself on camera? In the film industry, the term "frame" refers to how filmmakers position elements within a shot in relation to the edges of the screen. It entails selecting and arranging what is visible in the frame, directing the audience's focus and enriching the storytelling. Yes or no? If not, make a note under "Camera Framing."

☐

Camera Framing

_____

_____

_____

Do you have props and backdrops to spice up your head-shots or videos to establish your brand? Under "Props and Backdrops," write what you have, don't have, and would like to purchase.

☐

Props and Backdrops

_____

_____

_____

Let's shed some "light" on the situation — do you have ample lighting? If you own lighting equipment, take an inventory of what you have. Then, create a brief self-tape (in our industry, self-tape means video yourself) and assess whether your lighting is adequate. Do you look like you're in the dark? Are there overpowering shadows or light flares (light flares are when bright light hits a camera lens, causing visible artifacts and or spots)? Jot down your findings below under "Lighting." More on Lighting in Chapter 15.

☐

Lighting

_____

_____

_____

Assessing your sound system/microphone's functionality is of utmost importance. Do you own a microphone, and does it work well? To evaluate your sound quality, record the following video. Remember to articulate well, "Hello, I'm (your name). Have an amazing day!" Then, play back the self-tape and appraise the sound. Is it staticky, bouncy, low pitch, high pitch, hums, noisy, or crisp and clear? Take notes below under "Sound" to list any shortcomings.

☐

Sound

_____

_____

_____

Have you ever attempted to use a teleprompter? It's a super handy tool to have in your kit. There are lots of free options available online. If you use a teleprompter, write a short script and read from it while recording yourself. Do you gaze into the lens or fixate on the teleprompter? Do you appear natural, rigid, or just reading? Under the heading "Teleprompter," please jot down your findings below.

☐

Teleprompter

_____
_____
_____

What type of computer or laptop do you use? Determine if it's a PC or a Mac, as sound settings and editing tools vary (we'll discuss both sound and editing later). Write what you use under "PC or Mac" (desktop computer or laptop) below.

☐

PC or Mac

_____
_____
_____

Next, are you a video editing maestro? If so, what software do you use? Record it under "Editing." If you aren't familiar with video editing, write what training you need in this area.

☐
Editing

_____

_____

_____

Other Notes

_____

_____

_____

_____

_____

_____

_____

_____

_____

_____

_____

_____

_____

_____

_____

_____

_____

_____

_____

_____

_____

_____

_____

_____

_____

_____

_____

_____

_____

Now that your assessment is complete, you can dive into the most relevant chapters. However, I encourage you to explore every chapter where you may discover some golden nuggets. Without further ado, let's dive into "Unlocking the Perfect Hairstyle."

Chapter 3

Unlock the Perfect Hairstyle

Your hair is your personal magic touch. A great hairstyle frames your face beautifully, enhances your best features, and adds something special to your overall look.

Growing up, I disliked my hair. It was baby fine and challenging to manage, with the "static cling" hair flying all over, especially in the winter. My elementary school pictures were not flattering. I had yet to learn what hairstyle would flatter me. One day, I was excited as the curling iron entered my world in my early teens. My older female cousins (who were super cool) lived in Manhattan and knew what they were doing while I lived on Long Island in a small town called Hicksville. We saw each other often. One of my cousins, Daisy, was fun, beautiful, and stylish. I loved being around her; she made me feel special.

Daisy had a curling iron and showed me how to use it. She told me not to have the curling iron too close to my skin, but I burned my forehead a few times anyway. However, I kept at it and finally learned how to curl my hair without burning my forehead. Thanks to hairspray, my hair remained curly all day. I loved what I saw. It made me see myself differently. Finally, I loved my appearance and felt beautiful and confident.

I have had various hairdos throughout the decades. The 80s were out of the park with super high styles from teasing and puffing up. Working with my hair and knowing how to manage it was so much fun. It's a vital part of my existence. It was always worth the investment to try different things and see which styles made me feel fabulous — and the same goes for you too.

For fun, here are some hairstyles I crafted in the 1980s:

Your hairstyle should complement your face. It should enhance your best features, highlight your strengths, and create a harmonious balance. Think of it as building a masterpiece. Does it accentuate your eyes and flatter your cheekbones, jawline, and face shape? Does it divert attention from areas you may not want to emphasize? Hairstyles are not just about looks. They also tell a story. Your hairstyle reflects your personality and attitude, whether you prefer a sleek vibe, tousled charm, or a trendy new cut. It's a fun way to express yourself without saying a word.

Take a moment to pull back your hair (if needed) and study your face to figure out its shape. Find a celebrity with the same face shape as you to discover the best styles. Check out their hairstyles for some good ideas.

When choosing a hairstyle, you must consider your facial shape. You should also consider your personal preferences and the current fashion. Is your focus a professional look for work, a laid-back style, or something bold to make a statement? Try to choose a hairstyle that will fit many occasions.

Your hairstylist, especially if they are talented, is also an incredible resource. Seek their input on the best hairstyle for your facial features and personality. And remember — hair grows, so your style can always change.

Hairstyles can be one of the most dynamic aspects of personal expression. Exploring something new can make all the difference in your appearance and confidence.

If you're considering a hairstyle tweak, grab a pen and jot down your ideas. Remember to look on IMDb for celebrities who rock your face shape or use a "Hairstyle" App. It's your canvas, so sketch a few styles you want to try. Then, turn those thoughts into reality.

Next, we'll dive into makeup. Guys don't skip this chapter. It's for you too.

Notes

_____

_____

_____

_____

_____

_____

_____

_____

_____

_____

_____

Chapter 4

Makeup Magic for Everyone

Makeup is the magical art of enhancing your natural beauty. Whether you're an enthusiast or a novice, make-up is a tool for self-expression and creativity. Think of it as your artistic palette, allowing you to play with colors, textures, and styles to create a look that reflects your unique personality.

These days, I enjoy the ritual of applying makeup. It's like adding the finishing touches to my canvas. I love the tiny details, like matching my eyeshadow, blush, and lipstick to my outfit that day. I own twenty shades of eyeshadow, five shades of blush, and twelve different lipstick colors. You don't need to have this many, but your chosen colors will enhance your beauty.

Before applying makeup, check for any redness in your eyes. If you notice redness, 'Lumify' is an excellent redness-relieving eyedrop highly recommended by my TV news/sports anchor friend Rick Quan.

First, let's talk about the **Gal's** makeup.

Perfecting my makeup routine has been a decades-long journey. While your experience may be different, here is what works for me. After prepping my skin with serum and day cream (a chapter on its own), I start with concealer to even out my skin and cover any age spots and blemishes. Over that, I layer the foundation, blending it under my chin and neck. It's an important tip — you don't want a noticeable makeup line at the bottom of your chin or face. As an actor and makeup artist, I've noticed many films where the makeup artist neglected to do this, and it makes a difference — especially for older women.

Next, it's all about the eyes. I swear by a smudge-proof eyeliner pencil that gives me control over the thickness of the line. I apply black eyeliner to both my upper and lower eyelids. Then, on the lower lid, I press powder just below the liner to prevent it from running. For the upper lid, I tap a jet-black eyeshadow over the liner to keep it from smudging.

Sometimes, I'll use eyeliner underneath my eyelash lid for more depth. It is worth experimenting with — the key is to make your eyes pop. A little blending with the brush and some extra eyeshadow color add dimension. After perfecting my eye makeup, I darken my eyebrows with a thin brush dipped in jet-black eyeshadow. I'm not keen on using a pencil for my eyebrows because it looks unnatural. Makeup should enhance your beauty; not make you look made-up.

Now, on to the bronzer. I apply bronzer beneath my cheekbones, starting near the TMJ (temporomandibular joint, or the area in front of the ear) and gliding the brush toward the end of my lips. Then, I make a U-turn back up to the TMJ area to avoid heavy patches (repeat two times.) This method contours my cheeks, giving my face a slimmer appearance without leaving excessive bronzer near the lip area.

Next, I add blush, tapping off any excess on my brush before applying it to my cheeks for that natural look. Then, I apply pressed powder over my entire face to set everything in place, blending into the foundation, bronzer, and blush.

At last, the finishing touches — eyelashes, and lips. After a few curls with my eyelash curler, I add a couple of coats of "Thrive Liquid Lash" waterproof mascara. Finally, I apply lip liner and color my lips with either lipstick or lip gloss, matching the exact shade to my wardrobe. Then, I'm all set for my day and camera-ready.

Discover what shades complement your skin tone, experiment with different techniques, and find the perfect balance that enhances your features. But remember, there are no rules in makeup — only possibilities. Makeup is your chance to improve your beauty, express yourself, boost your confidence, and have fun along the way.

Now, Guys — it's your turn!

When you're on set or stage, posing for a headshot, self-taping (reminder: in the film industry, it means video-ing yourself), or presenting a live video on social media, it's common for guys to embrace makeup magic. Most people won't even realize you're wearing makeup, they'll only notice how great you look.

As a makeup artist, I've collaborated with countless men on set, and they always appear natural on screen. Here are some tips to look your best for the photoshoot, Zoom meeting, live stream, or role. Cleanse your face with a high-quality facial cleanser, followed by serum and facial cream on the night before and on the day of applying makeup. After a thorough cleansing routine, your skin will look and feel fantastic.

On the day of your production, start with concealer to cover up blemishes and age spots. Next, apply the foundation on your entire face, including forehead, and underneath your chin area. Then, add a light dusting of powder to achieve that flawless look. Remember to swipe lip balm afterward to keep your lips supple and smooth.

If you don't have full eyebrows, enhance them by applying a touch of eyeshadow or eyebrow mascara, using a color that matches your hair, if you have hair on your head. It's all about ensuring you enhance those brows and make them pop under the lights. You'll find many eyebrow mascaras and eyeshadows in the market-place. I recommend not using pencils to define brows since it looks unnatural.

For those of you with an oily complexion, keep some oil-free absorption sheets on hand to dab away any shine as the day progresses. You can buy oil-free absorption sheets on Amazon or at any store that sells makeup.

Here are some product suggestions for makeup. I recommend you visit a department store with multiple makeup counters, allowing you to go from one counter to another, such as Macy's, Nordstrom, or Lord and Taylor. You can choose any brand. Most TV people use MAC cosmetics. I'm not promoting Lancôme cosmetics; I'm simply sharing the Lancôme products I use as an example:

- Lancôme Teint Idole Ultra Wear All Over Full Coverage Concealer

- Lancôme Teint Idole Ultra Wear Buildable Full Coverage

- Lancôme Finish - Long-wearing - Full Coverage - Pressed Powder Formula

- Lancôme Color Design Single Eyeshadow Compact

Ask your salesclerk to recommend the proper brushes for any makeup you buy.

Whether you're into a natural glow or full-on glam, let makeup be your creative outlet and celebrate the beautiful canvas that is you.

Here are some pictures displaying various roles I've portrayed, each needing a distinct makeup look:

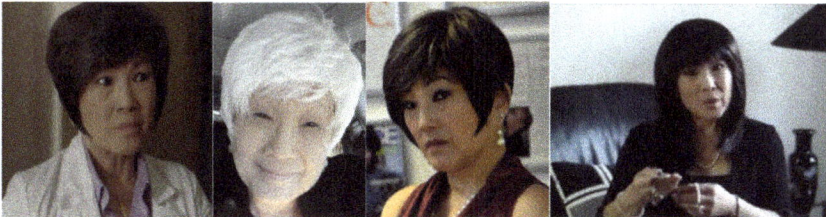

Need extra help? No problem!
Schedule a **FREE consultation** with Coach Judy using this link:
https://bit.ly/LCA30minAssessmentwJudy
or scan the QR code. Let's connect!

Until then, let's jump into "The Power of Your Smile" chapter.

Notes

Chapter 5

The Power of Your Smile

Did you know a simple smile is a gift you can give yourself and others without costing a thing? It's true. Smiling can brighten your mood and isn't just limited to you. Your smile can touch the lives of those around you, spreading happiness and joy everywhere you go.

I'm a big fan of smiling — it's a legacy in my family. I learned to smile while preparing for recitals from my ballet and tap classes. Over the years, my facial muscles have become seasoned with countless smiles at family events. Many included weddings and greeting customers in our family restaurant. Whether dining out, entertaining guests, or attending grand events in Manhattan like those at Radio City Music Hall, everyone expected me to maintain a cheerful demeanor. My husband of three decades once told me what first drew his attention to me was my smile. Smiling is a large part of my life, and I wouldn't have it any other way.

Do you love to smile? If so, I have a little tip for you. Don't just move your lips and cheeks; let your smile come from within and feel the joy and ease spread throughout your face. An authentic smile shows in your eyes, making your smile irresistible.

If it's not natural for you to smile, I encourage you to stand in front of a mirror and experiment with different smiles. Let your eyes show your inner happiness and find a smile that reflects your joy and vitality.

Taking care of your teeth is essential for both oral health and self-confidence. A joyful smile is always desirable. If you're unhappy with your smile because of your teeth, consider over-the-counter whitening options or talk to your dentist about your concerns. If needed, jot down notes on ways to enhance your smile.

Smiling is a powerful expression of positivity, so let's keep sharing it with others. In the next chapter, we'll tackle how to eliminate double chins.

Notes

Chapter 6

Banishing the Double Chin

Have you ever noticed that you have a double chin while posing and smiling in photos? It's a common concern for many people, regardless of weight. A double chin can appear when you smile or stand at an unfavorable angle, making you self-conscious.

Fortunately, there are simple ways to minimize a double chin in photos and daily routines. Try some of these tips and see what works best for you.

The next time you get your photo taken on camera, apply these quick steps:

- Stand or sit tall — imagine a string tied to the top of your head from heaven.

- Press your shoulders down.

- Push your shoulders back as far as possible.

- Bring your head forward, tilt your head a little to the left or right, and smile.

Practice these steps in front of a mirror and learn them well so they become natural when someone takes a picture of you, when you shoot a selfie, self-tape, or attend a Zoom meeting or webinar.

Here are some daily behind-the-scenes and off-camera routines to practice avoiding a double chin.

- Practice standing or sitting tall — imagine a string tied to the top of your head from heaven.

- Practice good posture, keeping your head tilted up and bringing it forward.

- Regularly perform neck exercises by vocalizing the letters 'Eeeeee' and 'Ohhhhhhh' to strengthen and tone your neck and jaw muscles.

- Drink plenty of water. It will help keep skin elasticity and reduce sagging under the chin.

- To perform chin lifts, tilt your head backward and look toward the ceiling. This helps stretch and tone the muscles under the chin.

- Consume a diet rich in fruits, vegetables, lean proteins, and whole grains.

- Be mindful of your posture when using electronic devices. Keep all devices at eye level to prevent tilting your head downward, which can contribute to neck sagging.

- Gentle massage under your chin helps stimulate blood flow and promote skin elasticity.

- When you apply a facial mask, be sure to include underneath your chin and neck for cleansing and a tightening effect.

These on and off-camera practices will help you avoid a double chin in your photos and videos. Take notes on what you need to do.

Next, let's discuss overcoming another significant confidence buster — wrinkles.

Notes

_____

_____

_____

_____

_____

_____

_____

_____

_____

_____

_____

_____

_____

_____

_____

_____

_____

_____

_____

_____

_____

_____

_____

_____

_____

Chapter 7

Combat Wrinkles with Confidence

We all experience aging. It's a natural part of life, but many would rather avoid discussing or acknowledging it. We cannot escape its effects on our faces, chins, necks, and other parts of our bodies. However, there's no need to worry about aging. You can handle it by caring for your skin. Proper skincare can help you look and feel your best.

Throughout my childhood, I watched my mother care for her skin. Through her example, I have gained a wealth of knowledge about skincare. Her dedication and commitment to caring for her skin have paid off — even in her mid-80s, her skin still is beautiful, almost wrinkle-free, and firm. My heart is full of gratitude for my mom's invaluable lessons and for how they have in-fluenced my skincare routine.

As a mother who values healthy skincare, I believe in starting early. I taught my daughter to care for her face from an early age starting in elementary school. Beginning with simple steps, like washing with a good facial cleans-er and moisturizing with a night cream, she learned the importance of hydrating her skin. Moisture is crucial in the battle against wrinkles, and I wanted her to establish good habits without delay. She can maintain a youthful glow by keeping her face clean, acne and pollutant-free. I'm proud to see her continue these practices into adulthood, and I hope she passes them down to future generations.

When people I meet discover my age, they often ask, "How do you maintain such a youthful appearance?" I answer them, I've cared for my skin for over 30 years with Origins products. I share my strict cleansing, moisturizing, and sunscreen routine. In addition, I use face masks and exfoliants to keep my skin looking fresh and rejuvenated, and I let them know I love using Origins natural products. Please understand that I am sharing my story and not advocating for Origins.

With the right products and routine, anyone can achieve healthy and moist skin. Remember that caring for your skin goes beyond maintaining your appearance and contributes to your overall health.

An early start is vital to healthy skin care, regardless of age. Individuals in their late teens and early 20s can begin by taking simple steps, such as washing their face with an excellent facial cleanser and moisturizing with a night cream and a day cream containing SPF 40. (Origins offers a night cream called Nightamins and day cream called A

Perfect World SPF 40 Age-Defense Moisturizer with White Tea.) Hydration is crucial for preventing wrinkles. You can support that youthful glow by keeping your face clean, acne-free, and pollutants-free.

As I have grown into my mid-60s, I have added various anti-aging products to my daily routine. I start my day with an anti-aging serum, day cream, and eye cream. In the evening, after I remove my makeup with a makeup remover and wash my face with an anti-aging cleanser, I use a spot removal serum, anti-aging serum, night cream, and eye cream.

From time to time, I allow myself a break from my regular regime to enjoy the "Origins Drink Up Intensive" overnight mask. Its texture resembles a moisturizer, ensuring it does not disrupt my sleep. My face, neck and upper chest get washed in the morning, and the overnight mask comes right off.

As you begin a new skincare regimen (adding a new cream or skin care product), it is essential to avoid skin irritation. By exploring, you can determine what is working for you. Read the ingredients, test serums, and facial creams on your wrist to see if you have an allergic reaction. If you do, ask a clerk to introduce you to other products.

I encourage you to begin with a night cream first that suits your skin type and concerns. Ask the makeup counter salesclerk to give you a sample. Once you have tested the cream on your wrist (and have had no allergic reaction), purchase it and use it for a week or two to ensure that your skin has no adverse reactions. Notice if you see a difference in your face. Are you glowing? Looking younger or refreshed? Also, remember to apply the cream to your chin, neck, chest (down to your cleavage, ladies), and the back of your ears (if you have a short hairstyle) to prevent wrinkles in those areas. Trust me, it works!

After using your first night cream for one month, you can introduce other products to your face, like a day cream, cleanser, toner, or serum (one at a time). This approach will give your skin time to adjust to each new product, and you can decide which products work best for you and which don't.

I use mineral, and detox masks to clean my pores from environmental pollutants. A mask once a week is excellent for this purpose. However, if it seems too much for you, start with once a month and increase the frequency. You will not regret it once you see the results of cleaner, smoother, and tighter skin.

Remember to listen to your skin and make changes as needed. If you notice any redness, itching, or other signs of irritation, stop using the product immediately and consult a dermatologist if necessary. With patience and care, you will find the perfect skincare routine to help you achieve healthy, youthful skin.

I encourage you to use natural products. I started with Origins Nightamins. It improved the appearance and texture of my skin overnight. I was so impressed.

It is essential to incorporate your skincare routine into your budget.

Exercising is another excellent way to tone and strengthen the muscles in your face, helping to reduce wrinkles. To enhance and tighten your facial muscles, try vocalizing "Eeeeee" and "Ohhhhhhh" (just as I mentioned in the prior chapter to reduce a double chin.) In addition, you can create resistance by opening your mouth and exerting pressure on the bottom of your chin with your thumb.

In addition, remember to take care of your lips. Applying your favorite lip balm before bedtime is an excellent way to moisturize them overnight. The lip balm will work longer and better since you won't be eating or drinking while asleep. If you're looking for a natural lip balm, "Burt's Bees" offers a variety of all-natural scented and flavored lip balms. I am not advocating for "Burt's Bees." It's just a suggestion.

Taking care of your skin is imperative for looking and feeling your best. Taking care of your face allows you to appreciate the beauty you have been blessed with.

If you'd like to know about all the Origins products I use, email me at judyg@lcaction.co.

Depending on your age and goals, list what you need in your notes to keep your face, chin, and neck moist and wrinkle-free. It's never too early or late to start. If you're a senior, consider using extra products with retinoids and ingredients that boost collagen. Remember that aging can be a wonderful experience, especially when glowing and looking fabulous. Write the goals you want to achieve and what you'll need to accomplish them.

Before moving on to the next chapter, it may be overwhelming if you're unfamiliar with taking care of your facial skin. I'd be glad to arrange a FREE personal Zoom consultation with you.

No need to **feel overwhelmed.**
Schedule a **FREE consultation** with Coach Judy using this link:
https://bit.ly/LCA30minAssessmentwJudy
or scan the QR code. Let's connect!

Now that you've taken care of your hair, smile, and face, you must present yourself with excellent posture. See you in the next chapter!

Notes

_____
_____
_____
_____
_____
_____
_____
_____
_____
_____

Chapter 8

Elevating Your Presence with Perfect Posture

Good posture is essential when presenting yourself — whether it's for a headshot, a conference, a Zoom meeting, a live video on social media, or getting that next gig. Besides increasing your confidence, good posture helps with balance, straightening your spine, and positioning your muscles correctly — it will make you feel great.

I wasn't born with good posture. As a child, my dad used to pull my shoulders back to prevent me from developing a hunched back. To this day, I appreciate my dad's efforts, yet the reason for my poor posture ran much more profound than a physical bent. I had low self-esteem. In elementary school, bullies abused me every day because, as a Chinese girl, I looked different from the Italian, Polish, and Jewish community. This daily taunting chipped away my self-worth and made me feel like I didn't belong. By shrinking within myself, I was protecting my heart.

In fifth grade, I found my voice in watching Bruce Lee movies every weekend in New York's Chinatown. I was tired of the bullying, and seeing Lee send his foes sprawling made me want to fight back. I responded to my bullies and stood up for myself. My verbal response shocked them, and they stopped bothering me. It was the beginning of a new life for me. I forgave my bullies and stood straighter.

I share this story in case it resonates with you. Is there anything in your life that makes you slump or withdraw inside yourself? If so, please record it in your notes to work through the emotional issues that may bring you down.

Note your posture at this moment. Are your shoulders and upper back slouching? If the answer is yes, here are some steps to help you improve your posture:

- When you stand, bring your shoulders back. It should push out your chest. Also, tuck in your tummy to give you a lift.

- Remember to bring your shoulders back while sitting. This should also make you push out your chest, so remember to tuck in your tummy for that extra lift.

- In every situation, whether it's gaming, reading, or driving, pay attention to your posture and keep your shoulders back.

If it sounds like I'm repeating myself, I am! I want this to be drilled into you because it helps.

When you work on your posture, it's normal to feel a bit of soreness in your shoulders, upper back, chest, and abdomen. It is a positive sign your muscles are stretching, tightening, and realigning themselves.

Strengthen those muscles by rolling your shoulders backward and forward. Do three sets of twelve reps, inhaling and exhaling, and you'll be well on your way to better posture in no time! Jot down your findings on your posture.

The next chapter is a fun topic — "Choosing Outfits that Pop on you."

Notes

_____

_____

_____

_____

_____

_____

_____

Chapter 9

Choose Outfits That Pop on You

Like most people, are you dreading clothes shopping? Many of us need help to find clothing that suits our style well.

A lack of knowledge about personal style can lead to dissatisfaction with our wardrobe. By understanding personal style, body type, and colors that complement our complexion, shopping can become more enjoyable and result in a wardrobe that feels both comfortable and stylish.

I struggled for many years to determine what clothes suited me. I used to just go with whatever style was in that season. Only when I reached my mid-twenties did I discover what looked good on me, which colors compliment my skin tone, and what styles to avoid. I went through a difficult learning curve where I tried stripes, flowers, and patterns, which made me look terrible. However, I realized solids were more flattering on me, and the best colors for my pigmentation and hair color were gemstone tones. Later, I discovered that the V-neckline is more flattering to my face shape than a U-neckline. If you still need to figure out what suits you, hang tight. It takes time.

Do you possess a wardrobe that you love and enjoy wearing? If not, do you know why you don't love your attire? There are many things to consider when buying the perfect tops, skirts, dresses, suits, shirts, ties, pants, patterns and styles.

You need to evaluate your body type. Are you tall or short? Do you have long or short legs? Do you have long or short arms? Be curious about your torso. Is it long or short? Are you thin, or do you have a muscled or chunky build? All these features matter when choosing your wardrobe.

Knowing what "Season" you are is very helpful. In fashion, seasonal tones are often associated with the colors worn during specific times of the year. It complements the natural surroundings and evokes the mood of the season.

In terms of the color palettes, let me give you a general overview of what is associated with each of the seasons.

The "Winter" tones are navy blue, charcoal gray, burgundy, emerald-green, and pristine white. Don't forget to consider the gorgeous jewel tones, such as sapphire, ruby, and amethyst.

The tones of "Spring" are soft pastels, which include gentle pink, baby blue, minty green, and lovely lilac. Springs can also rock those pops of brighter hues like coral, turquoise, and sunshine yellow.

The "Summer" tones comprise sky blue, sandy beige, crisp white, and soft peach. And let us not overlook those bold tropical colors like turquoise, coral, and lemon yellow.

In "Fall," the tones are a rusty orange, mustard yellow, olive green, and rich chocolate brown. To add a touch of elegance, incorporate jewel tones such as deep plum, emerald, and teal. So, whether it is winter, spring, summer, or fall, there is a perfect palette waiting to pair up with your skin tone.

To determine your season, try standing in front of a mirror and experimenting with holding pieces of your wardrobe up to your face. Ask yourself if those colors improve your face and eyes. Do you appear refreshed and youthful? Colors not in your season will make you look tired and washed out.

When determining if patterns and stripes are flattering, you rely on many diverse factors. If you have a bright, bold personality, you will look better in solids. If you are short (in height), avoid wearing horizontal stripes, which will make you look broader and shorter. Taller individuals with withdrawn personalities can benefit from wearing stripes and patterns such as paisley and flowers. You must try many outfits (tops, shirts, dresses, suits, and pants) to see what looks best on you.

A friend, Rick Quan, a former local news/sports anchor, found that solid-colored shirts in shades like white, cream, light blue, light pink, and lavender worked best for him. He noted that wearing a black shirt with a black jacket on air can create a "gangster" vibe, so it's best to avoid an all-black ensemble unless that's the image you're going for. He also stressed the importance of tailoring your suit and shirt for a perfect fit, rather than relying solely on off-the-rack pieces, which rarely suit individual body shapes. Tailoring ensures a polished, professional appearance, and adding a stylish tie for a pop of color exudes confidence while leaving a lasting impression.

To find what flatters you and makes you feel confident, you need to spend time in the changing room or return mailed items from Amazon. You may even want to seek fashion advice from a stylist, fashion coach or purchase the Color Me Beautiful book, by Carole Jackson. Here is the link to get her book on Amazon, https://bit.ly/ColorMeBeautifulbyCaroleJackson.

Without feeling restricted, dress in a way that suits your style, regardless of your age. To appear younger, I prefer not to dress in my age range. I enjoy looking youthful; it makes me feel great, and it brings me joy. Wear whatever makes you feel good about yourself! Feel free to make notes on any changes that may be needed.

In terms of youthful looks, we can now proceed to the next chapter, titled "Tips for a Youthful Appearance."

Notes

_____

_____

_____

_____

_____

_____

_____

_____

_____

_____

_____

_____

_____

Chapter 10

Tips for a Youthful Appearance

If you're wondering how to achieve a youthful appearance, the key lies in healthy lifestyle choices and taking care of your overall well-being. It means eating a balanced diet, sleeping well, staying hydrated, and exercising regularly. Protecting your skin from sun damage, pollution, and other environmental stressors can help support a youthful glow. Finally, prioritizing your mental and emotional health by managing stress, practicing self-care, and staying socially connected will keep you young inside and out.

This might feel overwhelming if you haven't already made these lifestyle choices. Take a deep breath and release any stress you may have. You won't be making all these changes at once, and you're not alone in this. I'm here to lend you, my support.

I can tell you I do them all. But understand it did not happen overnight. It's been a lifelong process for me. In my mid-twenties I wanted to be healthier, so I did the work and changed my life. I encourage you to apply the easiest changes first and incorporate them into your life so you can move toward a healthy lifestyle.

Here are some tips for healthy lifestyle choices that support your overall well-being.

Maintain a balanced diet of fruits, vegetables, whole grains, and lean proteins. Include foods with antioxidants, vitamins, and minerals for skin health. Work on staying away from deep-fried, greasy foods (the grease comes out on our skin.) Regular physical activity (taking walks, going for a run, or hiking) will improve blood circulation and promote a healthy complexion.

Aim for seven to nine hours of quality sleep each night to allow your skin to repair and regenerate. To avoid creating sleep lines and wrinkles, sleep on your back, not on the side of your face.

Drink plenty of water to keep your skin hydrated and support its elasticity. Use sunscreen with a high SPF to protect your skin from harmful UV rays. Many foundation makeup brands include sunscreen nowadays. Wear hats and sunglasses to shield your face from the sun (even use umbrellas if necessary).

Encourage yourself to commit to exercising three times per week, with a specific emphasis on enhancing both cardiovascular endurance and muscle strength. By incorporating regular exercise into your routine, you can increase your mood and energy levels, as well as maintain good health as you grow older.

My dad taught me the Chinese 312 exercise, a squatting technique. It was incredible to see him squat at his age. Every morning since 2017, I've incorporated this exercise into my routine. This method focuses on stimulating blood circulation, kidney and gallbladder function through targeted pressure on your hand and wrist, besides incorporating squats. Here's a YouTube link to learn more: https://bit.ly/Chinese312Exercise.

Set up a reliable skincare routine incorporating cleansers, moisturizers, and anti-aging products. Use products with retinoids and ingredients that boost collagen. Enhance your skincare routine with gentle exfoliation to remove dead skin cells and promote cell turnover.

Incorporate stress-relieving activities into your routine, such as meditation, yoga, or deep breathing exercises. Chronic stress has the potential to speed up premature aging. Develop a positive outlook to accept aging as a natural aspect of life.

Avoid smoking. If you smoke, work on quitting, as it speeds up the aging process and contributes to wrinkles and dull skin. Also, consume alcohol in moderation, as excessive alcohol intake can dehydrate the skin.

Dress in bright colors that complement your skin tone. This will give you a vibrant and youthful appearance.

Take a moment to write about the healthy lifestyle choices you are doing and are excited to practice. Trust me, you will feel the difference in your life soon enough — more energy, a brighter outlook, and much more confidence!

Boosting your confidence is what the next chapter is all about. See you there.

Notes

_____

_____

_____

_____

_____

_____

_____

_____

_____

_____

_____

Chapter 11

Radiating Confidence

Do you find it hard to feel confident in yourself? If that's true, know that you're not alone. Many people grapple with self-doubt and a lack of confidence.

You can gain self-confidence by using effective techniques. Self-improvement is attainable through methods shaped by psychology, behavior, and mindset. These strategies comprise self-affirmation, visualization exercises, and embracing a mindset focused on personal development. Remember that developing self-confidence is a gradual process that requires time and effort. By being persistent and open to personal growth, you can conquer self-doubt and reach your fullest potential.

Being introverted, confidence has always eluded me. It might be hard to believe if you've met me in person because I've trained myself to be a socially skilled introvert. I thrive at parties, conferences, and engaging with others. My journey started with ballet and tap, where I had to perform on stage, smiling and entertaining the audience. I felt scared deep down. I had to work hard to gain confidence, and I grew.

The Hilton Hotel in Orlando, FL hired me as a sales manager in my twenties in the hospitality industry. At a trade show, I had to speak about my property (to strangers much older than me) with no prior notice. I studied how they carried themselves and mimicked their professional postures and demeanors. I had overcome my fear and felt a newfound confidence in my abilities. When I returned home, I wrote positive reinforcements on Post-it notes, stuck them to my bathroom mirror, and repeated them to continue building my confidence.

Here are some strategies to enhance your confidence.

Repeating positive affirmations reinforces a positive self-image. You are fantastic and deserve to remind yourself of that each day. To do this, stand in front of a mirror and speak affirmations (from your heart) aloud while looking at yourself. Display on your mirror Post-it notes that contain positive words and phrases like these below:

- "I am beautiful."

- "I am wonderful."

- "I am attractive."

- "I am handsome."

- "I am creative."

- "I am talented."

- "I am intelligent."

- "I am equipped."

- "I am great."

- "I am a seasoned speaker."

- "I am confident."

Write whatever affirmations you need to hear to boost your self-esteem and confidence.

Here's a powerful exercise for you: close your eyes and visualize yourself thriving in different situations: speaking confidently to thousands at a conference, creating captivating and professional content, shining in front of the camera, lighting up every room you enter, and/or celebrating your book becoming an Amazon #1 bestseller. Feel the confidence and control as you embrace your success! Visualization helps you build a mental blueprint for success and reinforces your belief in yourself.

Practice confident body language through eye contact, maintaining a tall posture, and employing assertive gestures. In doing so, you will feel more optimistic. You will exude confidence and authority, and people will take notice.

Replace negative thoughts with positive self-talk and constructive affirmations every day. You can be your life coach and remind yourself of your strengths and capabilities. You are one of a kind, and you possess all the qualities necessary to thrive.

Take risks, leave your comfort zone, and face challenges head-on. Taking calculated risks can contribute to a sense of accomplishment and give you the confidence you need. Your bravery knows no bounds, and you have the power to conquer any challenge.

Stay away from negative people who criticize you and undermine you. Surround yourself with positive people who will uplift and support you. Positive social interactions can boost your self-esteem and reinforce your belief in yourself. You deserve to be surrounded by people who appreciate and value you. It applies to whom you allow on your social media platforms. Avoid being "friends" with people who denounce or bash others in their posts.

Practice mindfulness to stay present and manage anxious thoughts. Meditation can help develop a calm and centered mindset and reinforce your inner strength and resilience. You are in control of your thoughts and emotions, and you can maintain a positive outlook.

Learn to accept compliments and acknowledge positive feedback instead of deflecting or belittling yourself. You deserve appreciation and praise.

Be sure to exercise, even if it's not your favorite activity. I love what exercise does for my mind, heart, body, and spirit. Exercising improves my mood, self-esteem, and overall well-being. If you have not exercised before, start with finding a YouTube walking or cardio exercise video. You can achieve anything, especially with an exercise program that fits your lifestyle.

Also, wearing clothes that make you feel wonderful helps boost your self-assurance. You deserve to feel unique and confident in your skin as you wear the beautiful clothing you bought yourself.

Reflect on your achievements, strengths, and positive experiences. I started journaling in my twenties; it was very therapeutic. Journaling can reinforce positive thoughts and enhance your self-worth. Look at the remarkable things you've done. You should celebrate your successes. Consider achieving a "Success Board or Wall," as well. That way, when you look at your success board or wall, it will uplift your spirit. My board got overloaded over time, so I started a new one on the back of my office door. Seeing my achievements daily as I go in and out of my office brings me so much joy. You should be proud, yet humble, of all you have done.

Gratitude can also change your attitude. During painful, heavy, and hard days, I make it a point to stop what I am doing and count my blessings. Gratitude overwhelms us as we acknowledge the many reasons to be thankful. We deserve to feel amazing. Gratitude shifts our mindset towards positivity and enhances our assurance.

Building self-esteem and confidence is an ongoing process. Be sure to set realistic goals. Divide larger goals into smaller achievable tasks. Celebrate your successes along the way. You can achieve beautiful things, one step at a time.

Note the areas where you feel less confident and focus on improving your skills. Find what resonates most with you by trying different strategies, one at a time. Knowledge and competence contribute to self-assurance. Keep in mind that you are unlike anyone else and have the potential to accomplish what you set out to do.

You have what it takes to learn and grow, and you will become a shining expert in your field. Learn from your failures. Analyze your mistakes and use the experience to build resilience. View them as an opportunity for growth rather than a setback. Get some Post-it notes and write affirmations you need to see, hear, and believe.

Confidence radiates from within, inspiring and captivating your audience. In the upcoming chapter, we'll explore more techniques for captivating your audience.

Notes

_____

_____

_____

_____

_____

_____

_____

_____

Chapter 12

Captivating Your Audience

Preparation is the key to a successful presentation that captivates your audience. Write your speech and practice, practice, practice. This will help you feel confident, and it is contagious. When you're relaxed and delighted with your audience, they reciprocate.

Before you go in front of the camera or enter the building where you'll speak, do a vocal diction exercise by reciting all the vowels with excellent breath support, and say aloud to project A, E, I, O, U. Repeat three times and take a big breath in between each rep. I also say these phrases three times each in my vocal exercise before I self-tape or speak in person:

- A Large Appetite

- Big Black Bug

- Bubble Gum

- Chrome Appliances

- Cup of Coffee

- Eight Elephants

- I Love Ice Cream

- Incredible Fragrance

- Lazy River

- Orange Julius

- Saltwater Taffy

- Toy Boat

- Unique New York

Be sure to bring water and drink plenty of it to keep your vocal cords moist and ready. Also, consider carrying mints or pastilles to soothe your throat — I prefer Blackcurrant Grether's Pastilles. While they are pricey, they're worth it! I use them before speaking and for singing as well.

For a riveting opener, it helps to share a joke, a surprising fact, an interesting anecdote, or a thought-provoking question. I like to start with something funny. After being announced, I greet my audience with a big smile and say, "Yes, my name is Judy Go Wong. An easy way to remember my name is 'You Can't Go Wong with Judy!'" Then, I make drum sounds. Everyone always laughs; it is an excellent way for people to recall my name. Plus, everyone just laughed, and it breaks the ice in the room.

If I'm not the first speaker, I'll make everyone stand up and do stretching exercises like walking, jumping, lifting their arms, and massaging their neighbor's neck and shoulders. Also, I have them take a deep breath and roll their heads to loosen their necks. Then I invite them back into their seats, rejuvenated and ready to listen to me. I'll begin with a personal story to establish a connection.

Including visual appealing elements such as images, videos, or slides improves understanding and engagement. Attempt to personalize your message to align with the audience's challenges and concerns, regardless of whether you have delivered it previously. It's always good to spice up your presentation with a touch of humor. By adding humor, you can make your presentation more enjoyable, build connections and endorphins, and make it more memorable.

Secret tip: Speak while using your left eye to look into your audience's left eye. The left eye looks into people's souls, and this practice lets them see into yours as well. Building a genuine connection will be visible to others. While speaking, search for individuals in the room who show genuine interest by smiling and nodding. These people will encourage you, and you can build off their energy. Also, move around instead of staying stuck at the podium. Your body language exudes confidence as you navigate the stage, never needing to rely on the safety of the podium or the comfort of your notes.

Encourage audience interaction through polls, questions, or interactive activities. It will keep your audience invested in the conversation and make them appreciate doing something other than sitting and listening to you.

Remember that your audience wants to connect with an authentic, relatable person — you. Share personal stories, create common ground, relax, and have fun. It will influence your listeners and enhance their experience.

At the end of your presentation, invite questions from the audience to make your talk more engaging and address any queries they may have. You can also consider holding a raffle when they arrive to get them excited about winning something at the end. You can offer a free book, a QR code for a complimentary one-on-one session, or a link for a special rate to your high-value product or program.

Take notes on the strategies you want to use to captivate your audience. In the next chapter, we will discuss camera framing techniques.

Notes

_____

_____

_____

_____

Chapter 13

Frame It Right

What is camera framing when you are self-taping? It means positioning the subject (you) in the center of the frame (edges of the screen), from head to chest or waist. It is crucial for visual storytelling. Understand that the camera's placement can captivate and shape an audience's belief as much as the dialogue.

Filmmakers use various framing techniques to tell their stories, but for self-tapes, the most common is the medium shot (from the head to the chest or waist). The choice depends on your style and the message you want to convey.

As you frame yourself, consider your project's desired emotional impact, narrative emphasis, and cinematic style. Storytelling is a complex blend of visual and narrative elements. You're responsible for merging these techniques to create an engaging viewer experience.

I recommend using a PC or Mac's built-in camera for recording instead of investing in an expensive camera (but if you already have one, and if your screen can flip to show your frame and lighting, please use it.) I rely on my MacBook Pro and Zoom to record my podcasts, host webinars, and self-tape auditions. Mac has an outstanding camera on their computer screens and laptops.

If you use a laptop, have your screen in a straight position, and ensure the camera faces you at eye level. Using a medium shot, your head should be a half inch below the top of the frame, capturing your chest or waist to the bottom of the frame. If a standing desk is unavailable, prop your laptop on a box or enormous books to create the perfect height for the ideal frame. Or if you're using a desktop computer with a built-in camera on the screen, angle the screen at your eye level to avoid looking up into the camera. You should adjust the height of your seat, so the camera isn't above your eye level.

If you are using a video camera, place it on a tripod and flip the screen to view yourself. Position your camera at eye level and frame the shot with your head half an inch from the top of the frame and the bottom of the frame reaching your chest or waist. Keep your eyes on the camera while recording. Don't look at yourself on the screen. Keep your eyes glued to the camera lens on your video camera. If you keep your eyes watching yourself, it will appear like you are looking elsewhere. Remember, the goal is to connect with your audience and create a natural and flattering perspective. The same goes for using a camera, computer, or laptop.

When you set up your recording station, choose between standing or sitting. If you're going to sit down, use a stool without a back rather than a chair. Using a stool for filming yourself is ideal (unless you're an executive and wish to sit in your executive chair) as it won't be visible in the frame, especially if you use a virtual background. Before you click the record button, ensure you're in the frame, and check that everything is centered perfectly, including lighting, which we will cover in Chapter 15.

Be ready to document and gather the tools for your self-tape, whether you're standing or sitting, and for perfect framing.

Getting the perfect frame is one of many essential factors; having the ideal prop or background can add some spice to your video. We'll discuss that in the next chapter.

Notes

_____

_____

_____

_____

_____

Chapter 14

Enhancing Your Video Content

Any background is better than a messy room or a blank white wall for live meetings or self-taped auditions. Adding custom backgrounds to Zoom for a professional touch is easy.

First, pick any image you prefer as your background, whether it's your logo, or from Google images, or any app you have. Download the image onto your PC or Mac and label it for quick access.

Below are the step-by-step instructions on how to upload an image to your Zoom account.

Log in to your Zoom account.
Click "New Meeting" to enter the recording window.
At the top right corner see the recording and pause tab. Make sure it is on pause so that you are not recording yourself while setting up a background image.
At the top right corner, you'll see a green armor with a checkmark in the middle of it, known as the "Meeting Information" icon.

Click on the green armor, a window screen will pop up and you'll see a gear icon (at the top right corner of the screen.) Click on the gear icon, and another window screen will show up titled "Settings."

In the "Settings" window, on the left side, there is a list of topics.

Click the "Background & Effects," tab.

Select "Virtual backgrounds."

On the right side of "Virtual backgrounds," you'll see a + in the top right corner.

Click the + and a little window pops up and shows "Add image" and "Add video."

Select "Add image" and a new window of your download from your computer appears. You should see the image you wish to download.

Click on that image you want to use, then go to the bottom right corner and select "Open."

Your image will show up as your background in your Zoom frame.

Note that Zoom doesn't capture the entire uploaded image. The middle section of the image is primarily visible. Head back to the saved image in your downloads and crop the picture as you want it to appear on Zoom.

Once you've uploaded your cropped image and are happy with it, navigate to the top left corner and click on the red circle with the x to log out of "Background Images." To start your recording, locate the "Resume recording" tab in the top right corner of your Zoom screen by clicking on the greater arrowhead icon. If you're not prepared, click on the square icon labeled "Stop recording."

I have a vast collection of images in my Zoom virtual background file. I love reusing them.

If your recording space isn't ideal and you don't want to use a virtual background, consider buying a backdrop since cameras, video cameras, and smartphones don't have virtual background functionality, choose a background that complements your message.

On Amazon, you'll discover a range of options, from high-rise office windows to cozy fire living rooms with bookshelves, and rooms with solid colors. You can choose a personalized backdrop that highlights your logo or brand. If you decide to get a backdrop, you'll also have to get a backdrop stand.

A well-chosen prop can also enhance your next presentation, making it more memorable and engaging for your audience. From a beach ball to a hat, a mug to a rubber chicken, many fun options can help you convey your message effectively. Just be sure to use props in good taste and watch your audience's attention soar!

Here are some screenshots of my podcast and the backgrounds I've used.

Jot down your ideas on how you would like your video's background to look. Start by thinking about the mood you want to convey. What is your message, or what are you selling? If you haven't purchased an Image App, go to Google Images and type in "Free Images," or you can go to https://www.canva.com/photos/ to search for free images also. If you prefer a backdrop behind you, go to Amazon.com and look at the backdrops available for purchase. You may even want to design your own.

Notes

Chapter 15

Essential Lighting Tips

Good lighting is essential for filmmaking, video recording and headshots.

By shedding light on the subject, it improves the look of your video, highlights important details, and helps create a professional and polished result.

Throughout my time on House of Cards for multiple seasons, I have gained valuable insights into lighting. It is one of the most intricate aspects of filmmaking and it sets the mood.

It is best to use natural light when taking photos. Find a spot near a window or a well-lit area with diffused sunlight. In case you don't have access to natural light, you can use LED lights. You will need several of them to get the best results.

To light yourself, experiment with the three-point lighting setup (Key Light, Fill Light and Rim Light, also known as Back Light). Position the Key Light behind the camera, computer, or laptop on the right side to light up the subject's face. Place the Fill Light on the opposite side to reduce shadows on the left side. Station the Rim Light behind the subject (you) to increase depth and separate it from the background. You can project the Rim Light from the subject's top or bottom.

My Rim Light is on the bottom side of me, (on the floor behind my stool) to prevent a shine on my head. Because of the lack of natural light in my office studio, I use a small Ring Light (behind my laptop screen) in addition to the three-point lighting. Without it I look like I'm in a haze.

Be mindful of your surroundings, including the colors you wear and the background. Lighting is complex, so being flexible and acknowledging the need for adjustments is crucial.

Three-Point Lighting System Diagram

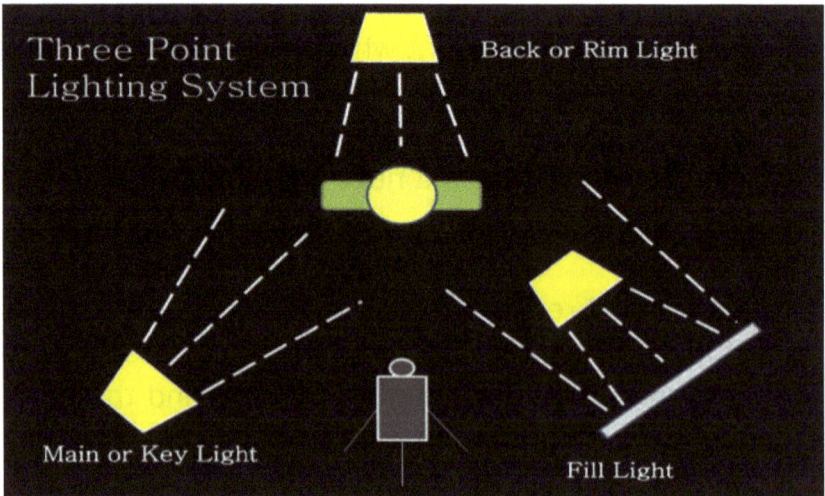

It is best to position yourself and your lights to minimize harsh shadows on your face, especially under the eyes, as they can be distracting.

By keeping the color temperature consistent across all your lights, you can avoid color imbalances. You can experiment with the color sheets that come with the LED lights. Pick lighting that suits your environment and be sure to adjust the light's brightness to prevent overexposure.

Here is a professional photography or video studio setup showing three-point lighting. The image includes a subject sitting on a stool in front of a neutral backdrop. The lighting comprises a Key Light on one side, a Fill Light on the opposite side, and a Rim Light positioned behind the subject, creating a balanced and well-lit scene. Include clear light stands and the soft box or reflector attachments on the lights.

You should explore different lighting angles to determine a setup that is both flattering and captivating. The tiniest modifications (a touch to the left or right) can affect the overall look.

Soft light sources such as soft boxes or diffusers produce illumination and gentle shadows, flattering facial features by minimizing harsh contrasts. Know that soft box lights will quickly raise the temperature in studios.

LED ring lights are a popular choice for video recording as they provide even illumination. You do not need to purchase expensive lighting equipment. White bed sheets or reflectors can bounce and diffuse light. If you use a Ring Light and wear glasses, you may need to adjust the angle of the light upwards towards the ceiling or tilt your glasses to reduce glare and reflections. I have a secret I want to share about wearing eyeglasses. I lower the brightness on the screen to eliminate the glare on my glasses. It works beautifully!

While many shapes and sizes of lights are available on the market, I recommend using LED lights. They come with color templates, and you can adjust the temperature and brightness with a switch for ideal lighting.

Below are images displaying several types and shapes of lighting. Explore them, alongside a wide range of other selections, on Amazon. Note your LED lights and list what else you may need. If you don't have any, you can compile a list of all the LED lights you want to purchase.

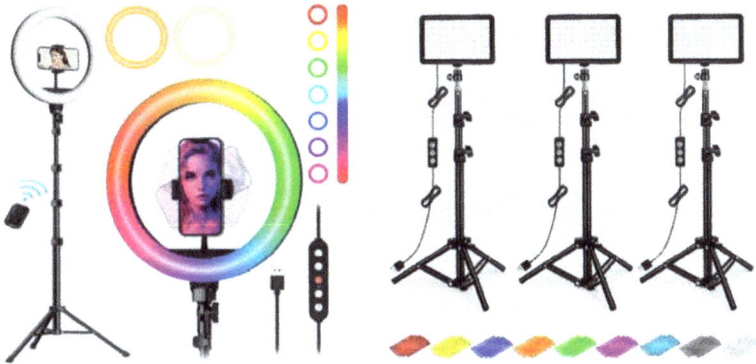

Sound is one of the most vital aspects of a successful video recording, next to lighting. Let's go to explore sound in the next chapter.

Notes

Chapter 16

Crystal Clear Sound

Sound is imperative to your video, podcast, webinar, online seminar, or meeting.

Most Macs and PCs come with convenient built-in microphones. But even with a good built-in microphone, you can improve your sound system. Dozens of great options are on the market.

External USB microphones connect to your laptop via the USB port and offer better sound quality than most laptop microphones. Some examples are the Blue Yeti, Audio-Technica AT2020USB+, and Rode NT-USB. I began with the Rode microphone for my auditions, then used an Audio-Technica microphone for my podcasts and coaching. Now I use my new MacBook Pro's sound system.

Lavalier microphones, also known as lapel microphones. They are popular among content creators and public speakers because they allow hands-free use. I own one and have used it several times. However, if you're an active speaker like me who moves around, it can pick up some unwanted rustling noises.

When choosing a microphone, consider factors such as sound quality, connectivity options, and budget. Reading reviews and comparing specifications can also help you find the best microphone to suit your needs.

I suggest purchasing multiple types of microphones on Amazon.com so you can test them. Document your experiments by recording yourself and noting which microphone produces the optimal sound. Return the microphones that don't meet your preferences.

Macs comes with a perfectly built-in microphone. If you own a Mac, contact me at judyg@lcaction.co for images of the settings on your Mac for crystal clear sound.

Once you have chosen the perfect tool for the best sound, you will be ready to plan your message.

Notes

_____

_____

_____

_____

_____

Chapter 17

Craft and Deliver Your On-Camera Message

To ensure that your on-camera presence is engaging and professional, it is crucial to plan your message before hitting the record tab. Your preparation will ensure that your screen presence is polished and that you come across as confident and knowledgeable.

The best planning process for me involves finding my key points, organizing my thoughts on what value I want to share, and creating a simple structure for my content.

First, name the title of the message you want to convey to your audience in a friendly, captivating, engaging, and approachable way. Think about what they need, what you want them to learn or understand from your video, and how you can best communicate that message.

Next, summarize the key points you plan to discuss in your video. Create a script or outline and integrate your message and key points. When you have finished articulating your primary thoughts, the message should seamlessly transition to a conclusion that includes a call to action, like for a program or coaching services they can join.

Remember to practice reading your message out loud to improve your delivery. You can also use a teleprompter or, for an even better result, memorize the script for a more natural presentation.

Record yourself on video, remembering to warm your voice for clear diction. Assessing the flow and delivery of your message will help you determine if it's outstanding. In conclusion, share your message sincerely and openly. It's easy for people to recognize authenticity.

Assess your performance in the self-tape. Is it remarkable? Would you sign up to work with you, or does it require some adjustments? Jot down areas that need improvement.

Now it is time to present your message and deal with those ever-present nerves.

Notes

_____

_____

_____

Chapter 18

Overcome Nerves for Success

So many of us know that feeling of butterflies when we are about to face the camera, speak in front of a crowd, or step onto a stage to sing or perform. Feeling jittery or unsure is normal, but guess what? There are some nifty tricks to help you handle those nerves like a pro and show up with confidence!

First, try some deep breathing exercises. (I recorded deep breathing exercises for my "Set Free First" clients; if you wish to try them, contact me at judyg@lcation.co.) It's amazing how taking a few deep breaths before jumping into the spotlight can calm those nerves and help you feel more centered and composed.

You can also try visualizing your success. Imagine yourself owning the moment and rocking whatever you are about to share. This is a fantastic way to boost your confidence and kick your nervous feelings to the curb.

Let me share a little secret about how I tackle those nerve-racking moments. What supports me when speaking on stage, on camera, or gearing up to sing with all my heart? Thorough preparation! Familiarizing myself with my material ahead of time helps me feel more at ease. Perfect practice makes perfect!

I do vocal warm-ups to ensure excellent diction for speaking gigs and podcasts. I polish my script until it is superb and rehearse like no tomorrow. When it is time to sing with the worship band at church, I practice on my own two weeks prior, so the band practice is excellent. You better believe I'm up at 4 a.m. on the days I serve with my church's Worship Band, warming up my voice for those practice sessions.

Through lots of practice and sticking to a routine, I have boosted my confidence and squashed those fears of messing up. Sure, I have had hiccups, but I have learned that being prepared and focused is the key to feeling confident and ready for anything.

So, if you're ready to conquer those nerves and shine bright on camera or stage, just remember to take a deep breath, visualize success, practice, practice, practice, and most importantly, have fun connecting with your audience!

Write down a routine and what you must do to conquer those nerves.

Oh, and those of you who are focused on self-taping, keep an eye out for my homemade teleprompter — it's been a meaningful change for managing nerves on screen. Let's dive into that in our next chapter.

Notes

_____

_____

_____

_____

_____

_____

_____

_____

_____

_____

_____

_____

_____

_____

_____

_____

_____

Chapter 19

Create Your Script Assistant

If you want an easy and cost-effective teleprompter, I recommend using Microsoft 365 OneNote. I have experimented with several types of teleprompters, including a physical teleprompter setup that used a camcorder and an iPad to read my script.

Although many teleprompter apps are now available for computers, they cannot solve a vital issue: when you read from the teleprompter, are your eyes focus on the words rather than the camera, making it obvious to your audience that you're simply reading to them.

Let me tell you why Microsoft 365 OneNote is the ultimate homemade teleprompter for me. With OneNote, the words I'm reading from my script are at the top of the screen (right under the word OneNote) near my camera lens. I click on Home to close the top bar, so there is not much seen on the top of the OneNote window. I slide the window up on my screen as high as I can, where the word OneNote is aligning with the camera lens. Then, I'm able to read my lines and appear to be addressing my audience without looking like I'm reading from a script. It is the perfect tool to help me deliver a confident and flawless performance.

When using OneNote, I narrow my script to see only three words per line. This reduces eyestrain and improves readability. It also looks pleasing to my audience because it will **not** look like my eyes are "reading from a script" by moving back and forth across the screen. And my eyes appear to be looking right at my audience (because my left eye is on the word OneNote) connecting with them.

I also use my trackpad to control the speed of my script. Most teleprompter apps have a predetermined speed you can choose, but with OneNote, I can pause for effect and adjust my speaking pace without the added stress of keeping up with scrolling words.

However, choose the tool that best suits your needs. Take notes on what you have tried, and which product or app works best for you. You must practice, practice and practice. Then, record yourself reading from a teleprompter and assess your self-tape. Do you look professional addressing your audience and are you captivating, or do you look like you're reading from a teleprompter?

Once you have successfully recorded yourself, it is time for some simple editing!

Notes

Chapter 20

Editing Excellence

Editing is a bit of movie magic. It can be fun yet *very* time-consuming. However, if you primarily record webinars and self-tapes, you only need to know a few editing basics.

If you use a Mac, iMovie is a user-friendly software with many bells and whistles. If you use a PC, the Microsoft, free editing software is Clip Champ, it's also easy to use. Many people also like using Adobe and YouTube Studio. Pick the one that works best for you.

There is no need to spend money on expensive editing software unless you require advanced editing tools. It depends on your preference and your desired features.

If you are creating a self-tape, all you need to do is clip the beginning and the end. No one needs to see the moments before you speak, and no one needs to know when you're looking for the record button to stop. That's all you'll need to cut and delete unwanted footage.

Everything I mention in the next few paragraphs is from iMovie. Clip Champ doesn't come with a lot of bells and whistles unless you pay to upgrade.

Depending on your lighting, you may also adjust the contrast, which changes the brightness or darkness of your video clip. You can experiment with it to create the proper effect for your recording.

The same goes for the sound. You can adjust the volume if you think you are flat, too loud, or too soft.

If you think your video's speed is too slow, you can increase it. If you think you are speaking too quickly, you can also slow down the speed. Then, there is also "fading or transition" at the beginning and the end. You may want your video to fade to black in your opening shot and at the end. In iMovie, there are a lot of transitions to choose from for the opening shot and the end.

Are you considering music at the beginning, throughout, or at the end of your video? First, decide on the music and download it into a file. Then, click on Audio and Video and upload the desired music to your music file. Drag it to the timeline and location where you want the music to play.

Should you want to add a logo at the beginning or end of your video, click the arrow, (Import Media), find your logo image and click on it, then go to the bottom right corner and click Import Selected.

Pick your favorite editing tool and practice using it. It's that simple! Learning these tools takes time, so be patient with yourself. You'll eventually learn them, practice using them, and be a pro at them. Jot down what you want to use and what areas you need help with.

You're nearly at the end of this workbook! As we head into the closing chapter, I have some personal insights to share with you.

Notes

_____

_____

_____

_____

_____

_____

_____

_____

_____

_____

_____

Chapter 21

From Judy's Heart: A Personal Note to You

Congratulations! You have made it through all the pages of this workbook. I hope you have found insights and advice you can use throughout your life and career.

Now, let's take a moment to reflect on what you need to do next to achieve your goals. List your priorities and figure out a few areas where you'd like to start. Then, take a moment to find the top three topics that require your attention and prioritize those based on importance. That way, you can focus on the areas that will most help you achieve your goals.

If you feel overwhelmed, remember that you don't have to go through this process alone. It's incredibly helpful to seek support and another person's perspective. Don't hesitate to ask for a second opinion or to have someone by your side through your journey.

If you want personalized solutions, I'll be glad to help. Remember, you can set up a free 30-minute consultation with me. So be sure to write all your thoughts and questions before we get together.

THANK YOU VERY MUCH FOR TAKING THE TIME TO READ
"LIGHTS, CAMERA, ACTION! MASTER YOUR ON-CAMERA PRESENCE LIKE A PRO." I AM EAGER TO HEAR YOUR FEEDBACK.
PLEASE GO TO HTTPS://BIT.LY/LCAWORKBOOKFEEDBACK OR TO THE QR CODE TO SHARE YOUR THOUGHTS.

I would greatly appreciate it if you would write
a review on Amazon.com.
Here is a link and a QR code.
https://bit.ly/WriteaLightsCameraActionReviewonAmazon

Schedule a FREE, 30-minute session as soon as possible to discuss your aspirations and how Coach Judy can help you with your needs and/or answer your questions at https://bit.ly/ LCA30minAssessmentwJudy or scan this QR Code.

I look forward to seeing you shine with your radiant smile when you master your on-camera presence like a pro!

If you have questions, reach out to Judy Go Wong at judyg@lcaction.co.

www.ingramcontent.com/pod-product-compliance
Lightning Source LLC
Chambersburg PA
CBHW071649210326
41597CB00017B/2158